**CONTENTS**

# INTRODUCTION

As America's armed forces face an extremely complex and chaotic strategic environment, the ability to maintain an advantage over an uncertain adversary is a constant challenge. The world is currently seeing an increasing number of economic and political superpowers, a rise in the ability of state and non-state actors to produce nuclear weapons and an increase in the capability of violent extremist organizations to fund and support terrorist activities. During this time of global insecurity, the United States (US) must stay determined and focused in order to protect the American people, its allies and partners. In order to accomplish this task, the US currently relies heavily on technological means which are becoming rapidly the predominant choice for defeating global threats. From the organizational structure to the facilities utilized for training, warfighters have become significantly dependent on technology and the capabilities innovation provides. Unfortunately, the negative effects created due to this over-reliance result in significant dangers to the warfighters that often go unrecognized. The purpose of this paper will be to identify the negative effects created due to the armed forces over-reliance on technology and to provide recommendations for current and future incorporation of these innovations.

Technological innovations have provided unprecedented capabilities throughout history, repeatedly giving military forces a marked advantage over the enemy. In the distant past, the development of gunpowder, rifled muskets, iron clad warships, railroads and the telegraph revolutionized the conduct of war. More recently, the atomic bomb, intricate global positioning systems and pinpoint precision guided munitions are sophisticated capabilities that have dramatically changed the strategic, operational and tactical levels of warfare. All of these innovations have proven successful on the

1

battlefield and could be considered the critical factor contributing to victory. And as

American military forces remain engaged in global conflicts and also prepare for future

engagements, the demand for more advanced capabilities will continue to thrive.

To the detriment of current service members, an over-reliance on these

technological advancements has created dangers and negative implications that are often

unrecognized or ignored. It is suggested that "modern technology inhibits military

performance, gets in the way of individual effectiveness, thwarts initiative, and detracts

from skill on the battlefield."[1] In an article discussing Technology and Logistics it is

stated that "The real issue is to recognize that technology is a tool with limitations, and

these limitations should be considered in reacting to particular situations. Technology

does not offer a silver bullet for all situations."[2] Despite the common understanding that

technology has limitations, Service members too often rely on these capabilities as a

solution and are blindly allured by the sophisticated appeal of innovative equipment.

Regardless of the dangers an over-reliance on technology creates, the reality is

that technology plays an integral role in modern warfare. It is a force multiplier and will

continue to enhance battlefield capabilities. The challenge is to properly integrate

technology without conforming to an over-reliance despite concerns related to manpower

cuts, resource constraints, budget restrictions and over-stretched forces. Unfortunately, to

counter these restrictions military forces tend to revert to technology as a remedy, further

enhancing the existing over-reliance. The relevance of this topic is that technology

broadly impacts American armed forces across every spectrum of warfare, and if the

---

[1] William J. Holland Jr., "Technology is Key to the Operation Art, Not an Obstacle," *Proceedings,* Vol. 130, Issue 4 (April 2004): 2.

[2] The Editors, "Technology and Logistics," *Air Force Journal of Logistics, Vol. XXX, Number 1,* (Date Spring 2006): 45.

negative implications associated with ineffective implementation of technology are not considered, an unnecessary risk must be accepted in the methods in which Services conduct war and the inevitable dangers to individual Service members.

## Thesis

As American armed forces strive to maintain a distinct advantage over uncertain global adversaries in a complex, uncertain and resource constrained environment, it is common for warfighters to depend on technology as a solution. These technological solutions often create hazards and vulnerabilities that either directly or indirectly undermine and compromise the safety and combat effectiveness of military forces. However, advanced technology continues to be introduced to all levels of warfare at an alarming rate. This paper argues that American military forces have developed an over-reliance on technology that lures warfighters into ignoring the dangers associated with misapplied innovation.

## DOTMLPF Assessment

To defend this thesis the author will organize supporting arguments by the commonly understood acronym DOTMLPF (doctrine, organization, training, materiel, leadership, personnel and facilities). This method provides a logical and cohesive format for identifying areas of concern that essentially span all aspects of military operations. Each chapter will separately identify a relative issue, provide arguments to support the thesis and lean towards recommendations. Recommendations will be primarily provided in a separate chapter. The facilities portion of DOTMLPF assessment will not be included due to lack of substantial supporting arguments and limited information on the effect technology has on physical locations.

**CHAPTER 1: DOCTRINE**

Doctrine has contributed significantly to military successes throughout history, but doctrine has a modern competitor that challenges its fundamental guidance on the battlefield. This competitor is technology. And as the US armed forces rapidly become more sophisticated, an over-reliance on the capabilities technology provides continues to increase. The two goals of this chapter are to identify the effects of this over-reliance by illustrating the necessity of a healthy relationship between technology and doctrine, and to highlight commanders' tendencies to allow technology to outpace doctrinal guidance. To refine this argument, the author will address three areas: the definition of doctrine, the importance of effective doctrinal interpretation and implementation, and the effects created due to an existing over-reliance on technology.

Joint Publication 1-02 defines doctrine as the, "fundamental principles by which the military forces or elements thereof guide their actions in support of national objectives."[1] Essentially, doctrine is the way the armed forces conduct business. In simple terms, doctrine is a play book that outlines the methods in which units man, train, equip and sustain forces while identifying the tactics, techniques and procedures to accomplish any number of missions.

General Curtis Lemay emphasizes the importance of doctrine as he states,

At the very heart of warfare lies doctrine. It represents the central beliefs for waging war in order to achieve victory. Doctrine is of the mind, a network of faith and knowledge reinforced by experience which lays the pattern for the utilization of men, equipment, and tactics. It is the building material for strategy. It is fundamental to sound judgment.[2]

Despite the importance of doctrine identified above, Lieutenant Colonel (LtCOL) Paul D. Berg, US Air Force Chief of Professional Journals, mentions in an article on doctrine and technology that Service members "view doctrine as being akin to broccoli–

---

[1] U.S. Department of Defense. *Dictionary of Military and Associated Terms*, Joint Publication 1-02 (Washington, DC: Secretary of Defense, April 12, 2001), 166.

[2] U.S. Secretary of the Air Force. *Air Force Doctrine Document 1*,(Washington DC: Secretary of the Air Force, November 17, 2003), 1.

healthy but boring."[3]  And in reference to technology, Lt Col Berg goes on to say, "eyes are apt to glaze over at the mere mention of the word *doctrine,* yet *technology* conjures up exciting visions of sophisticated equipment."[4]  There is no doubt that technology is exciting and if implemented correctly can greatly assist in achieving victory in almost every conflict.  But put into context, technology only provides the means (resources) to accomplish tasks, not the ways (methods).  For this reason, the existing over-reliance on technology overshadows the significance of doctrine and has a tendency to become the predominant choice for accomplishing military objectives.  And as technology continues to advance and provide unmatched capabilities, the concern is that "there is always the danger that technology will make one's doctrine obsolete, will replace doctrine as the determinant of the future, and will become merely a convenient shibboleth endowed by advocates with greater significance than it in reality possesses."[5]

What must not be overlooked is that both technology and doctrine must maintain a healthy relationship that is critical to success.  Lt Col Berg emphasizes this relationship as he states, "doctrine deserves respect because it has been intertwined with technology for millennia and explains how to use technological tools to achieve military purposes. Skillfully integrated doctrine and technology can lead to victory, but technology without doctrine to guide its use has little military significance."[6]  Dr. Hallion expounds on this relationship by saying, "Neither is independent of the other; rather, each generates a

---

[3] Paul D. Berg, "Doctrine and Technology," *Air & Space Power Journal* (Spring 2008):  20.

[4] Ibid, 20.

[5] Richard D. Hallion, "Doctrine, Technology, and Air Warfare:  A Late Twentieth – Century Perspective," *Airpower.au.af.mil.*  Fall 1987. http://www.airpower.au.af.mil/AIRCHRONICLES/apj/apj87/fal87/hallion.html (accessed February 24, 2011).

[6] Berg, 20.

synergistic impulse that encourages and strengthens the other. The lagging of one is necessarily injurious to the other."[7] In summary of these supporting statements, both doctrine and advanced technology must coexist in a give and take relationship, though one cannot function effectively without the other. In a perfect world, they will both have the flexibility to adapt and evolve based on the current and future military requirements, while considering the effects on the other. This marriage is important to understand as it identifies a conditional relationship that frames the support for this paper's thesis.

The author has identified several conditions regarding this relationship between technology and doctrine. The first condition exists when technical advances outpace doctrine, where technology is employed with minimal direction. The second condition is when doctrine outpaces technology and the technical capabilities do not exist to accomplish the doctrinal guidance. The third condition is when doctrine and technology are intertwined too closely together. Understanding that all of these conditions create various effects, the goal in this argument is to stress that the current over-reliance on technology tends to advance without doctrinal support. This tendency masks the importance of doctrine and causes commanders to redirect efforts towards the utilization of technical capabilities to develop new concepts for the methods of war. The focus of this paper will be to highlight the issues pertaining to the first condition when technology outpaces doctrine.

When technology outpaces doctrine several negative effects result. First, without doctrine to guide the implementation of technological advancements, the result often leads to the development of fanciful and often irrelevant technical capabilities. Second,

---

[7] Hallion.

with the availability of elaborate technology and no doctrine to support its use, commanders often find themselves inadequately educated and trained to properly employ these sophisticated technologies. The final result of this condition is attempting to change the conduct and methods of war based on technical capabilities. These conditions will be illustrated in several historic engagements, and in turn, highlight the necessity to manage the current over-reliance on technology.

As an example of the first negative result, the Nazi's developed the "blitzkrieg" during World War II, which consisted of the incorporation of infantry, mechanized units and aerial support to deliver an overwhelming combined force against its adversaries. Having placed extreme confidence in this operational capability, the Nazi's had planned for a short, decisive victory, and did not anticipate a long war. Due to this assumption, doctrine remained stagnant and the technical research and development groups were left without operational guidance. In Dr. Hallion's words, "this led to technological fanciful projects more related to World War III than World War II projects such as ballistic missiles (a wasteful drain on the German research and development and war economy effort)."[8] Had Germany focused on the development of applicable doctrine to guide the proper implementation of existing technology, the results of WWII may have been different. In this case, technology outpaced doctrine and created a vulnerability with devastating results.

Despite this historic lesson, American military forces have recently made similar mistakes. For example, the Army invested heavily into the multi-billion dollar Future

---

[8] Hallion.

Combat Systems (FCS) program[9] which was terminated due to cost increases,

production delays, and failed technology. This capability was, as Ezio Bonsignore

suggests,

> Once touted as the US Army's 'centerpiece modernization effort,' and indeed it encompassed an extremely ambitious plan–actually far too ambitious, as it eventually became painfully evident–to leverage on far reaching advances in virtually all aspect of land warfare technologies to radically change not only the way the US Army is equipped, but indeed also its structure.[10]

The result of this over-reliance on technology stemmed from a fanciful vision that

delivered no new operational system, utilized valuable funds and kept personnel from

better supporting the Service members currently deployed. Had the over-reliance on

technology been weighed against doctrine that illustrated relevant operational guidance,

the failed attempt at the unrealistic program may have been prevented.

To illustrate the ineffective use of advanced technology by commanders that do

not have proper doctrine to support it and who do not have a true understanding of how to

utilize the advances, Dr. Hallion discusses the 1917 British stalemate in Cambrai during

World War I. He states that the use of technologically advanced tanks had "promise of

converting the existing war of stalemate into a war of movement, perhaps resulting in a

decisive breakthrough of Allied forces."[11] However, due to the lack of artillery and

infantry support, the tank offensive was brought to a halt. Dr. Hallion continues to say,

"the introduction of a technologically superior weapon-the tank–had been frustrated by

---

[9] The FCS program was developed to provide a technologically advanced C4I system that incorporated the latest information technology, an integrated command and control network and state-of-the art equipment consisting of Manned Ground Systems, and Unmanned Air and Ground Vehicles. All of these systems were networked together to provide the best situational awareness and the most advanced tactical warfighting capabilities.

[10] Ezio Bonsignore, "A Technology Too Far: "Reviewing" (or Terminating?) the FCS Programme," *Military Technology* (July 2009): 12.

[11] Hallion.

total lack of appreciation of how to use and support such a weapon."[12]   What could have been a decisive capability for the British resulted in mediocre results due to the over-reliance on technology in conjunction with the lack of supporting doctrine and the understanding of how to utilize artillery and infantry support.  If the over-reliance on the tank was not seen as the single method for victory, appropriate steps may have been taken to reinforce existing tactics, techniques and procedures.  Had the appropriate doctrine been developed to determine how to effectively use the tanks, the outcome could have potentially been more favorable and the advantage of the Allies improved.

The over-reliance on technology without the marriage of doctrine was also apparent with the development of precision-guided munitions (PGM)[13].  This innovative creation led leaders to believe that wars could be won with technology alone and defeat the will of the enemy to fight with the strategic targeting of high value targets.  With the use of PGMs, Christopher Schnaubelt states, "Following Operation Allied Force in 1999, some analysts argued that the campaign over Kosovo demonstrated the capability and joint combined airpower to force enemy capitulation without the need for boots on the ground."[14]   This supports the thought that technology when utilized with precision targeting and effects based operations would all but eliminate the need for ground troops, in an effort to reduce costs, collateral damage and friendly casualties.  And with the recent memories of the Vietnam War still fresh in the minds of many military leaders, the reluctance to commit a large force appeared to logically support the intent.

---

[12] Hallion.

[13] Precision-guided munitions (PGMs):  Munitions guided through global positioning systems, either ground or aerially guided.  PGMs offer the flexibility for precision targeting with extreme lethality due to the extreme accuracy with only a potential for fractional error.  PGMs also represent an advantage in terms of stand off capabilities.

[14] Christopher M. Schnaubelt, "Whither the RMA?" *Parameters* (Autumn 2007):  97.

Dr. Steven Canby of the International Institute for Strategic Studies explains this phenomenon as he states, "the current explosion in technological solutions to military problems and the associated development of doctrine, is being termed the most recent 'Revolution in Military Affairs (RMA)'," and adds that "new technologies only affect the techniques, not the nature or principles of war."[15] This supports the argument that technology tended to drive doctrine without considering the current strategic environment and the enemy American forces were engaging. The first implementation of this new RMA philosophy occurred during Operation DESERT STORM/DESERT SHIELD (ODS/DS) that proved to be successful based on the mission tasking. Donald Rumsfeld then capitalized on the opportunity to leverage this technical experience to transition the force to a Rapid Decisive Operational (RDO) organization. Unfortunately, this technological revolution proved successful during (ODS/DS) which only reinforced the policy supporting the movement and skewed commanders' basic current philosophy on the fundamentals of warfare. This success spawned an epidemic confidence in technology that urged commanders to maximize precision targeting and to minimize troop allocation. In turn, this revolution continued to narrow the focus of research and development into innovations that supported the technological RMA while quickly creating a separation from the doctrine that did not support these types of operations.

With this same philosophy in mind the US entered Operation ENDURING FREEDOM (OEF) and Operation IRAQI FREEDOM (OIF). Both of these campaigns were initially efforts of rapid victory through the reliance on RDOs and PGMs with little regard for the real aims. To support this philosophy Adam Herbert adds, "With an

_____

[15] Steven L. Canby, *New Conventional Force Technology and the NATO-Warsaw Pact Balance, New Technology and Western Security Policy* (Basingstoke, UK: Macmillan 1985): 66.

unparalleled ability to detect enemy forces rapidly and deliver precision munitions against high-value targets throughout the depth of the battlespace, U.S. forces were expected to decisively outmatch any potential adversary and fully dominate every contest."[16] Unfortunately, this dependence resulted in insufficient ground troops being allocated for the ensuing war, which is highlighted by Christopher Schnaubelt in an article discussing the technological RMA theory as he states, "the minimal size of ground forces deployed and available for OIF was the result of planning to fight the war we envisioned, with RMA-capabilities we hoped for, instead of the enemy and conditions we would actually face.[17] If technology such as PGMs would not have carried such overwhelming emphasis in the assumptions for a rapid and decisive victory in OEF and OIF, planners may have recognized the type of campaigns American forces were about to enter. The over-reliance on technology led planners to believe that the two campaigns could be won by technological means based on the advantage over the adversary. In reference to this belief, Major D. Williams states, "Nations must accept the risk associated with the extensive employment of technology, but, in so doing, may jeopardize the human element essential to war fighting."[18]

To summarize, emphasis should be placed on understanding the context of doctrine and the relationship doctrine has with technology. This relationship should remain dynamic and allowed the freedom for change. More specific to this paper's hypothesis, it is important to ensure technology does not outstrip doctrine. If it is allowed,

---

[16] Adam J. Herbert, "Army Change, Air Force Change," *Air Force Magazine, 89* (March 2006): 37.

[17] Schnaubelt, 95.

[18] Major D. Williams, "Is the West's Reliance on Technology the Panacea for Future Conflict or its Achilles Heel?" *Defense Studies*, Vol.1 No.2 (Summer 2001): 40.

negative effects such as the ones discussed will be created. These negative effects can be prevented if several precautions are implemented and the relationship between doctrine and technology is understood.

# CHAPTER 2: ORGANIZATION

An organization has several definitions, "Something made up of elements with varied functions that contribute to the whole and collective functions, a group of persons organized for a particular purpose, or a structure through which individuals cooperate systematically to conduct business."[1] All of these definitions apply to the military context of an organization; however, the topic of discussion in this chapter is not aimed at the organization itself. The focus will be on the organizational impacts that the implementation of advanced technology create in the areas of command and control and span of control.

When faced with the challenge of improving the efficiency and effectiveness of an organization, a commander has multiple options. They can refer to typical methods such as the implementation of rigorous standard operating procedures, the execution of numerous battle drills or the conduct of repeated training exercises. All of these methods are valid choices, but technology has added a new dimension to the capabilities a commander has at his fingertips. With these capabilities, a commander can retrieve immediate information to speed the decision making process, calculate coordinates at the push of a button and receive clear satellite communications from anywhere in the world; all of which contribute towards efforts to improve the organization of a command. In most cases these advances add significant capabilities to an organization, but negative effects and vulnerabilities often result if not implemented correctly or managed properly.

---

[1] Farlex, "The Free Dictionary," http://www.thefreedictionary.com/organization, (accessed March 9, 2011).

## Centralized Command and Control

As today's military faces manpower reductions, resource constraints and increasing global requirements, commanders are faced with the task of creating an organization capable of meeting any challenge within the designated scope of responsibility. In order to accomplish this task, a great focus of effort has been placed on improving the efficiency, effectiveness and timely response of command and control (C2) processes, a major component of every command organization. To improve this area, decision makers turned to technology as the solution.

A recent attempt to utilize technology to improve organizational effectiveness was illustrated through the Army's Force XXI concept. Steven Komarow explains that the Army "Has developed a taste for technology, as seen in the systems that support Force XXI and Army XXI concepts," in an effort to "use technology to streamline its command hierarchy and field lighter, more lethal forces for the twenty-first century."[2] The key to Steven Kamarows' statement is the effort afforded to streamline the organizational command structure. Unfortunately, this streamlined hierarchy can cause several conditions that result in negative effects.

A common condition when seeking a streamlined chain of command is the adaptation of a vertical, or centralized, command structure. This structure meets the objective of improved efficiency, but at a cost. This vertical structure forces an

---

[2] Steven Komarow, "Army Forces to See Major Restructuring," *USA Today*, February 16, 1999, A1. During the era of a technological revolution in military affairs, Force XXI and Army XXI were concepts that incorporated the effects of integrating technology into preexisting areas to include doctrinal flexibility, strategic mobility, tailorability and modularity, joint, multinational, and interagency connectivity and versatility in war and Operations Other Than War (OOTW). The Force XXI concept, along with the technological revolution of military affairs, was the cornerstone in the twenty first century for incorporating technology into currently existing doctrine to create advantageous effects. This concept seemed to mark the beginning of a major dependence on technology and the results technological advances could potentially have on military operations.

14

organization to minimize the number of subordinates within the decision making process, and; whereas, in a flat (decentralized) command structure more individuals have leadership positions and are encouraged to engage in more independent decision-making. The vulnerability with a technically enhanced centralized command structure is a commander's temptation to exercise centralized control, which can potentially counter any technological efforts to improve efficiency.

Until the turn of the century, tactical and operational commanders had the direct authority to make decisions during daily events, exercises or tactical activities. Unfortunately, current technological advances provide the means for a commander to minimize subordinate leadership responsibilities. Michael Skaggs paraphrases S.L.A. Marshall in reference to World War II as he states, "many U.S. company commanders in the Pacific were under constant pressure from headquarters to report information. Worse yet, they were often ordered to take tactical action based on the headquarters' estimates of the situation."[3] These decisions were often made without a realistic picture of the tactical environment. It is apparent that the more information a commander has available, the more apt they are to make lower level decisions. John Gentry explains, "Enhanced connectivity already increases the general's ability to micromanage tactical operations from afar – a propensity they have demonstrated repeatedly with lesser communications capabilities – to the detriment of initiative, operational performance, and the morale and retention of junior officers."[4] This is also enforced by Skaggs when he states, "A headquarters' perception of the situation provided by command and control PC

---

[3] Michael D. Skaggs, "Digital command and control: Cyber leash or maneuver warfare facilitator?" *Marine Corps Gazette* 87, no.6 (2003): 46.

[4] John A. Gentry, "Doomed to Fail: America's Blind Faith in Military Technology," *Parameters* (Winter 2002-2003): 100.

(personal computer) is a threat to the tactical commander whose immersion in the situation is far more accurate. That perception can, in many cases, become the catalyst for the abandonment of decentralized control and in turn lead to a tactical defeat."[5]

Now, consider the battlefield effects based on the intricate C2 systems of today and consider that the potential for this to take place with the current network centric environment. Skaggs goes on to say, "In a modern military, equipped with command and control technology such as Blue Force Tracker, centralized control is more likely."[6] However, John Carozza advises, "The operational commander and his staff must exercise restraint by permitting decentralized execution of his COA [course of action] by tactical commanders."[7] In essence, the technology that was intended to provide a commander with a common operating picture now provides the commander the ability to micro-manage subordinate leaders.

### Increased Span of Control

The best strategy is always to *be very strong*, first in general, and then at the decisive point. Apart from the effort needed to create military strength, which does not always emanate from the general, there is no higher and simpler law of strategy than that of keeping *one's forces concentrated*. No force should ever be detached from the main body unless the need is definite and *urgent*. [8]

---

[5] R.L. Taylor, Technology Sound not Technology Bound: The risks of over-reliance on modern military capabilities, (Quantico, VA, Command and Staff College, February 19, 2009), 7.

[6] Skaggs, 46.

[7] John L. Carozza, *The Unspoken Consequence of Command, Control, Communications Technology: Enhanced Micromanagement by Risk-Averse Commanders*, (Newport, R.I.: Naval War College, February 9, 2004), 15.

[8] Carl Von Clausewitz, On War, *On War*, trans. Peter Paret and Michael Howard (Princeton, NJ: Princeton University Press, 1986), 240.

As the geographic responsibilities of a commander increase, they are forced to develop concepts that support growing global requirements. To meet these increasing demands, the answer tends to result in a distribution of forces across an area of responsibility. This is often referred to as span of control. A span of control may vary depending on current requirements. At times, it may be very insignificant with only a few satellite elements in the field or it may consist of units spread over a vast geographic area in multiple locations. And as forces are strategically positioned to source theater engagements, security cooperation activities and campaigns, it is essential for commanders to ensure that control is maintained between every unit.

The ability to maintain control of these vastly dispersed elements is only possible due to advanced technological capabilities. Robert Bateman states, "Current command, control and communications (C3) technology permits the operational commander to lead from remote locations while maintaining a near-real-time understanding and control of the battlespace."[9] This luxury gives commanders the choice where their command and control centers will be geographically located, which removes distance as a restraint when leading units from afar. Commanders now have the option to lead troops in combat from thousands of miles away, sometimes without any personal experience on the battlefield. John Carozza states, "this separation can turn the conflict into a 'video game' where the consequences of ineffective decisions are diluted to such a point that they are no longer recognized by the commander or his advisors."[10] Regardless of the negative results, some commanders remain compelled to retain operational control without a true

---

[9] Robert L. Bateman, *Digital War: A View From the Front Lines* (California, Presidio Press, 1999), 17.

[10] Carozza, 11.

17

picture of the situation on the ground.

An advanced technical capability that has been extremely effective at persuading commanders to control from remote locations is the video-teleconference (VTC). During Operation ANACONDA, one of the first major combined operations, during OEF, video-teleconferencing played a major role in the miscommunication between commanders. James McPherson states, "The over-reliance on VTCs in Operation Anaconda at the expense of personal face-to-face interaction degraded General Franks' ability to gauge his subordinates' level of understanding. Component staffs came to rely solely on VTCs for coordination to the exclusion of message traffic."[11] According to Dr. Milan Vego, the elements involved for planning and executing the operation were located around the world in a virtual C2 consisting of

> The Combined Forces Land Component Commander (CFLCC) in Kuwait, the Combined Forces Air Component Commander (CFACC) in Saudi Arabia, Joint Special Operations Task Force (JSOTF) Dagger in Karshi Khanabad (K2), and JSOTF K-Bar in Kandahar, CJTF MTN, created for Anaconda by CFLCC, started out in K2 and moved to Bagram in mid February 2002.[12]

Dr. Vego goes on to say, "Evidence suggests that the collaborative planning between components for Anaconda was accomplished almost exclusively through VTC."

Despite the apparent advantages a VTC may provide, it is common during a conference session that many unnoticeable implications occur. McPherson says, "Differing perspectives seem to be common in VTCs and may be too subtle for participants to notice. A traditional face-to-face meeting between key members of the

---

[11] James A. McPherson, *Operation Anaconda: Command and Control through VTC*, (Newport, R.I.: Naval War College, 14 February, 2005), ii.

[12] Milan N. Vego, "Operational Command and Control in the Information Age, " *Joint Forces Quarterly 35* (2004): 103.

component staffs early in the planning cycle might have revealed this difference in perspective."[13] Mark Davis says during his study of Operation Anaconda that "VTCs not only eroded the interpersonal relationships between component commanders and component staffs, but they also eroded General Franks' ability to lead. General Franks' decision not to forward deploy forced a heavy reliance on VTCs to coordinate operations with his component commanders."[14] In many senses, Operation Anaconda was deemed a complete operational failure due to the lack of communication and coordination. The operation could have been a great opportunity to begin OEF with a success; however, the over-reliance on technology, such as the VTC, to close the gap between time and space for coordination resulted in a less than favorable conclusion.

As advanced communication technology leads to new and improved methods for streamlining command efficiency, an over-reliance on these capabilities can cause units to self-impose organizational changes that result in degraded organizational effectiveness. Caution must be used to refrain from over extending forces and straining command and control relationships based on the mere availability of advanced technology. Control and good leadership decisions must be in place to instill restraints on this over-reliance.

---

[13] James A. McPherson, 8.

[14] Mark G. Davis, "Operation Anaconda: Command and Confusion in Joint Warfare" (master's thesis, School of Advanced Air and Space Studies, Air University, Maxwell Air Force Base, Alabama, 2004), 12-13.

# CHAPTER 3:  TRAINING

As the operational and tactical environment becomes more complex, providing the most effective training becomes increasingly challenging.  J.D. Fletcher states, "Limitations in time, funding, training devices, training personnel, ranges, supplies, and other resources such as ammunition and fuel have made training missions increasingly difficult to accomplish."[1]  To overcome these limitations and to provide the most feasible training opportunities, technology often becomes the source for the solution. This chapter will discuss the use of technology as a tool to enhance current training methods and identify several vulnerable areas created in the process.  These vulnerabilities include the availability of time to train, the effectiveness of the current training priorities and the lack of sufficient technological training resources.

## Impact of Operational Tempo (OPTEMPO)

As the United States now finds itself ten years into the overseas contingency operations, military forces have felt a strain on personnel as a result of lengthy and repetitive deployment cycles.  This is often referred to as OPTEMPO, which is measured by the time a Service member spends away from his home station.  Over the past ten years, deployment cycles have been a tremendous strain on both military organizations and individual Service members.  This demand for forces overseas has often resulted in the deployment of units for a significant portion of their service obligations.  Joint Vision 2010 states, "The capability to control the tempo of operations and, if necessary, sustain a tempo faster than the enemy's will also help enable

---

[1] J.D. Fletcher, *Virtual Reality:  Training's Future?* Edited by Seidel and Chatelier (Plenum Press, New York, 1997), 169.

our forces to seize and maintain the initiative during military operations."[2] This reference suggests that deployment rates will not decrease, but continue to increase in the near future. With a reduction in time available for training, leaders must make critical decisions as to what pre-deployment training milestones and objectives should take priority.               During inter-deployment phases commanders must decide what training objectives will be met. In the current complex environment, units have more objectives to meet than they have time to meet them. In an article discussing eroding Marine Corps navigation skills, Trista Talton states, "Commanders ultimately decide which skill to drill during pre-deployment training and which one to ignore. The strain of seven month-on, seven month-off cycles means commanders have to prioritize, so sometimes basic skills are taken for granted."[3] Ultimately, each commander has the responsibility to properly train his units to meet their tasked goals and objectives, and it is up to those commanders to make the most of their time. This forces leaders to abandon time intensive fundamental training, and strive to qualify an operator to know just enough to get the job done. Unfortunately, "As equipment becomes more complex, more time is required to learn how to operate and maintain it, leaving less time to learn why it works, how to question whether or not it is working correctly, and how to continue to the mission without it."[4] However, as long as technology can provide the necessary capability, it is easy for a commander to bypass in-depth, individual training objectives.

---

[2] Chairman of the Joint Chiefs of Staff (CJCS), *Joint Vision 2010,* (Washington, DC: Government Printing Office, June 2000), 15.

[3] Trista Talton, "Corps tackles erosion of navigation skills," *Marinecorpstimes.com.* June 7, 2009. http://www.marinecorpstimes.com/news/2009/06/marine_land_nav_gps_060709w/ (accessed February 24, 2011).

[4] David R. Price, *Technology In Transformation: Critical Strength or Critical Vulnerability,* (Newport, RI: Naval War College, May 18, 2001), 12.

If leaders do not have to invest extended time training to basic skills, they can quickly meet training requirements and move on to other priorities. This creates a vulnerability that may not give operators the necessary skills for mission success.

**If Technology Fails: Abandoning Basic Skills Training**

When applying technology to a situation, one must determine how much risk he is willing to accept if that technology were to fail. In order to make that determination, one must study the associated vulnerabilities to determine tolerable levels of risk and cost for that application.[5]

Technology will eventually fail, but do soldiers have the basic warfighting skills to operate without it? And how do commanders justify the reliance on technology to substitute for fundamental skills training? Currently, commanders face the difficult challenge of preparing subordinates for real world operations. Whether the Service members are on the front lines or in a joint operating center, they must be prepared to handle any situation, and the level and amount of training each Service member receives is a direct reflection on how they will perform under pressure. Unfortunately, technology is often used as a crutch to provide capabilities while proper and effective training is sacrificed.

For the purpose of this discussion, the author will focus on land navigation and the dependence on the Global Positioning System (GPS). In this example, Talton summarizes the current reliance on the use of technology as she states, "Land nav is

---

[5] David Hoey and Paul Benshoof, *Civil GPS Systems and Potential Vulnerabilities*, (Eglin Air Force Base, FL: Air Armament Center, October 25, 2005), 5.

especially vulnerable because of the easy access to technology as a solution."[6] In turn, military forces have become over reliant on GPS for navigation and often ignore basic skills such as map reading, terrain studies or orientation. In reference to these basic skills, Raymond Millen states, "the general attitude is that these devices have made such training unnecessary."[7] Trista Talton adds, the "Marines are now so dependent on the devices–some say addicted–that many are at risk of disaster if they lose power or signals in critical combat situations."[8] Sgt. Damian Senerchiea found himself dependent on a GPS in Hit, Iraq where he states, "We were using our GPS as a compass, it put us in a bad position, especially in that urban environment. We didn't know where we were."[9] Despite the slight chance that a GPS will fail, the consequences are potentially disastrous if warfighters are not taught to manually navigate without the aid of technology. Again, a commander must weigh the benefits of dedicating valuable training time towards a more comprehensive understanding of the traditional training principles, and resist the false sense of security that technology provides a fail proof capability.

To prevent this scenario, Michael McPherson recommends that, "The services should train their forces to use GPS-dependent systems to be able to operate in a GPS hostile environment, or one in which GPS is not available. Without it, the armed forces could experience an over-reliance on GPS and a degradation of basic navigational skills."[10] No experience is greater than that of live and realistic training. Through the

---

[6] Talton.

[7] Raymond Millen, "The Art of Land Navigation GPS Has Not Made Planning Obsolete," *Infantry Magazine*, January-April 2000, 36.

[8] Talton, "Corps tackles erosion of navigation skills," *Marinecorpstimes.com.*

[9] Ibid..

[10] Michael R. McPherson, *GPS and the Joint Force Commander: Critical Asset, Critical Vulnerability*, (Newport, RI: Naval War College, May 18, 2001): 15.

realism, commanders can test their subordinates, force them to be as uncomfortable as possible during and expose them to an unlimited array of training challenges. Training should be difficult, so when the troops get on the ground they are fully prepared and are capable of fighting in situations without the reliance on technology.

A more devastating event relates to the failure of the GPS network when associated with precision guided munitions (PGMs). For example, "In late 2001 and early 2002, errant US munitions killed allied troops and many Afghan civilians."[11] The GPS system in this situation failed resulting in multiple, unnecessary deaths. Michael McPherson suggests that the Armed Forces should,

> Not rely on GPS as the sole navigational, precision guidance weapons guidance, and timing information in military weapons systems and command and control systems. Over-reliance on GPS produces critical vulnerabilities in the operational concepts of precision engagement, force protection, operational maneuver, and command and control."[12]

With an over-reliance on these capabilities, the risks and vulnerabilities must be weighed against the benefits of the capability. Commanders must understand that the potential for failure exists and must take every precaution towards implementing critical steps to prevent the failure and ensure multiple capabilities are in place to back-up existing innovations.

### Simulation Training: Not a Perfect Substitute

In order to maximize individual training experiences while minimizing the logistical investment of people, funding and time, the military has extensively invested in automated training simulators. To many, these virtual training systems have become the

---

[11] John A. Gentry, "Doomed to Fail: America's Blind Faith in Military Technology," *Parameters* (Winter 2002-2003): 90.

[12] McPherson, *GPS and the Joint Force Commander: Critical Asset, Critical Vulnerability*, ii.

ideal training environment. They are produced in a variety of configurations ranging from laptops systems that test a Service member's ability to lead troops in combat, to state-of-the art flight simulators that test the skills of the most qualified aviation crews. Despite the advantages, and the perceived realistic training these simulators offer, many negative aspects result from training with an automated system.

In many cases, operators gain a sense of indestructibility when facing an automated simulation. Just as if an individual were playing a video game at home, there is no recourse for poor decisions. No physical or emotional connection can be simulated when a life threatening situation occurs. The bullets, the casualties, the fear for life and the camaraderie cannot be simulated. The operator has no physical comprehension of pain, extreme environmental temperatures or exhaustion. This realistic experience can only be gained during live training or in combat. In this case, commanders must ensure they do not completely depend on simulator training as the primary source of instruction. Even leaders find the simulator training useful, they must also incorporate realistic training to fully test the knowledge of the warfighters. Without this combination, the loss of realism is a vulnerability that operators may not overcome when faced with real world situations.

As warfighters prepare for future engagements and technological capabilities become more elaborate, it is essential that basic training principles not be abandoned. In order to prepare the most effective fighting forces, leaders must ensure that appropriate priorities are placed on fundamental training skills regardless of the challenges and obstacles. These training priorities should be the focus of effort to best prepare warfighters for the complexity of the security environment and for the safety of the

Service members.  As leaders, the focus of effective training is not only to meet training requirements, but to provide the fundamental tools to be able to defend themselves on the battlefield.

# CHAPTER 4: MATERIEL

American military forces are equipped with the most high-tech, innovative materiel[1] in the world. From individual weapons systems to unmanned aerial platforms, military forces have increasingly used technology to gain a marked advantage over the enemy. And as the armed forces have become increasingly reliant on these materials they often find themselves unable to reach mission success without the aid of technology. This dependence, in combination with the extensive inadequacies of the acquisition process, creates a risk for the Service members that may not be overcome.

Every soldier, airman and sailor expects the equipment they use to operate correctly, provide accurate information and deliver rapid results. Whether an intelligence analyst expects an information collection system to produce a priority of targets, or a soldier is depending on a global positioning system (GPS) device to deliver them to an exact location, Service members demand results from the technical capabilities they have available. Unfortunately, there are material factors that must be considered before depending solely on technology to provide required capabilities. In order to address these issues, this chapter will discuss the current military acquisition process, the effects high costs have on the ability to provide sufficient equipment to the warfighters and the lack of interoperability associated with the acquisition of redundant capabilities. This chapter does not necessarily focus on technology as the problem; however, the argument will focus on the over-reliance on technology that creates a demand for innovative equipment

---

[1] JP 1-02, 227. Materiel: All items (including ships, tanks, self-propelled weapons, aircraft, etc., and related spares, repair parts, and support equipment, but excluding real property, installations, and utilities) necessary to equip, operate, maintain, and support military activities without distinction as to its application for administrative or combat purposes.

regardless the negative effects such as the deployment of immature technologies and equipment that may not be interoperable with existing systems.

## Acquisition

The acquisition of materiel is not a simple process. Contracts, money, procedures, delivery, stovepipe production,[2] quality assurance and operability are a few of the many areas of concern when dealing with acquisition process. However, it does not excuse the fact that the warfighters need materials delivered quickly that are functional and meet the demanded requirements. The 2010 Quadrennial Defense Review (QDR) states that the Department of Defense (DoD) should be able to "rapidly create prototypes and field new capabilities, maximizing its ability to meet warfighter needs and leverage technological advantages."[3] Despite great efforts to accomplish this task, the current acquisition does not meet this requirement. The speed of the current acquisition process is not effective for delivering timely materials. And in one of the most demanding technological fields, information technology, the delivery process is having extreme difficulty keeping up with demand. According to the 2010 QDR, "The Department must reform the acquisition process with a view to accelerating the acquisition cycle."[4]

When attempting to accelerate the acquisition process, "Another obstacle for military systems is the Defense Department's 'stovepipe' approach to developing

---

[2] Stovepipe production refers to different vendors, organizations, and military units that produce separate parts for one system. At any point, a delay in the production of one part could increase the delivery time of the entire system.

[3] US Department of Defense. *Quadrennial Defense Review Report*, (Washington DC: Government Printing Office, February 2010), 93.

[4] 2010 QDR Report, 93-94.

technologies."[5]  For example, when discussing the current necessity for satellite communications, "the Pentagon's traditional approach to buying hardware is unhelpful for systems such as sat-com networks, which require the integration of components made by different vendors."[6]  This stovepipe process is often slowed when one or more parts are not produced simultaneously with the completion of other parts, which compounds the delay in delivering end products.  This also results in technical difficulties when the parts are assembled.  One part may not operate properly, which results in additional delays when a single contracting company must make the corrections or repairs.  The development of this is process is only a product of the over-reliance on technology and the increased demand for technical capabilities.

An additional issue related to the slow acquisition process is the intricacy and technical parts required to produce fully operational equipment.  For example, the sophistication associated with satellite communications is overwhelming.  It is estimated that to create a Tactical Satellite (TACSAT) system to support the current global communications requirements, "would take 10-15 years to build.  Such a long development cycle raises the questions of whether the money should be spent on more mature technologies, even if they are less sophisticated."[7]  Many of these advanced capabilities are based on fanciful projects that lead to the production of equipment too expensive to produce or employ, ending in the cancellation of projects due to the inability to meet time and budget limitations.  Cancelling these programs result in gaps in

---

[5] Sandra I. Erwin, "Ground Connections:  Mobile Broadband for Roaming Troops:  Pipe Dream or Reality?" *National Defense* (June 2008):  49.

[6] Ibid, 48.

[7] Ibid, 49.

production, lost time and wasted expenditures. Had the focus of effort been on realistic programs, the outcomes may have been more advantageous. This forces warfighters to search for alternative means to fill capability gaps. In this case, the viable option is commercially available substitutes that may not be as elaborate or sophisticated, but equally meet the required capability.

The obvious question is, how does a slow acquisition process negatively affect the situation on the battlefield? In order to answer this question, it must be understood that the demand on the battlefield drives supply. In other words, basic economic principles control how much, how fast, and how often products are developed and pushed off the production line to the warfighter. Along with the increased dependence on technology, the warfighter demands more advanced materials even more rapidly. In addition, this equipment is expected to perform without flaws and solve every intended problem. Unfortunately, warfighters are the victims of their own success. And as long as a reliance on technology exists, Service members will expect the availability of technical capabilities to support the demand. However, the control of this dependency lies in the hands of the commander. With operational and tactical insight, direction and a focus on fundamental skills, the demand can be minimized to a controllable level.

## Immature Technology

In many cases, the high demand for technology places a rush on the delivery of products, ultimately resulting in the employment of immature technology. This often presents inadequate performance of materiel, operational failure during combat and unfortunately, in some cases, the loss of American lives. This could be prevented if the technology was properly tested, analyzed and screened prior to sending the equipment in the field.  However, "the maturation development can be expected to delay the fielding of a new system or upgrade for a few years."[8]  Based on the immediate demand for technical capabilities and the necessity for rapid production, vendors are not always afforded sufficient time to perform the necessary measures to ensure equipment is fully functional prior to delivery.  In many cases, most of the testing takes place when the systems are utilized by the Service members in the field.  In this situation, the vulnerabilities are exponential to the safety of Service members when untested equipment is employed on the battlefield.  If the new technology fails or falls short of intended expectations, it is essential that commanders have multi-layered systems that back up new technology with similar equipment that can perform the task required.

An additional concern for immature technologies is that they often lack the logistical and maintenance support in order to sustain the equipment in the field.  Without this in place technologies that fail do not have replacements, spare parts or the technicians qualified to repair the technology.  As warfighters become dependent on these technologies, it is detrimental to mission accomplishment if they have not considered back-up plans or cannot conduct operations until a replacement arrives or the existing equipment is repaired.

## Costs

Technology is expensive, and as American armed forces face extreme budget cuts over the next few years, the ability to purchase technological advancements will become more difficult.  With the increased reliance on technology and a decrease in available funding, how will forces be able to meet the demand for technical capabilities?  Several

---

[8] John Dumond and others, *Maturing Weapon Systems for Improved Availability at Lower Costs,* (Santa Monica, CA:  RAND, 1994), xiii.

issues exist when discussing the latest advances. Walter A. Vanderbeek argues that "Emerging technology is expensive. These expenses tend to force the development and implementation of technology to take on a life of its own, which at times can run contrary to the most efficient use of the systems being developed. It is very possible to develop a system that costs too much to be allowed to be threatened in combat."[9] Dawn Onley states, "System costs prevent everyone from having the latest technologies."[10] She continues to state that the Blue Force Tracker or the Force XXI Battle Command Brigade and Below (FBCB2) communications system "costs about $10,000 per ground unit to install."[11] This makes it difficult to fund the sourcing for every unit requiring the capability. In addition, "the Defense Department has spent billions of dollars on mobile satellite-based communications to bring connectivity to forward deployed troops. But the available systems are scarce and too expensive for the military services to be able to field in large quantities."[12] At this cost, Services cannot afford to outfit every warfighter, vehicle, operations center, or command with the latest innovations. In this case, insufficient supply does not meet the demand. This negatively impacts end users when, as the example of the FBCB2, the majority of the Army elements greatly depend on this navigational equipment and use this tool as a lifeline for mission success.

---

[9] Walter A. Vanderbeek, Technology or Doctrine, Search for Balance, Core Course Paper (Fort Leslie, WA: National War College, 1994), 7-8.

[10] Dawn S. Onley, "Systems help avert friendly fire deaths," *gcn.com.* May 16, 2003. http://gcn.com/Articles/2003/07/24/Systems-help-avert-friendly-fire-deaths.aspx (accessed February 24, 2011).

[11] Onley.

[12] Erwin.

## Adapting to Demand

As forces seek immediate technological solutions, the military often utilizes the commercial-off-the-shelf (COTS) process to acquire materials. In many cases, this program also encountered the slow and cumbersome processes as the traditional acquisition methods currently used. On the other hand, the Army has taken steps towards increasing the speed of acquisition through alternative means to meet capability requirements and overwhelming demand. This process is called the Army's Rapid Equipping Force, a spin-off of the COTS process and is, "revolutionizing the way the Service gets new technology into the hands of warfighters."[13] Donna Miles continues to say, "The Rapid Equipping Force concept is the traditional military acquisition system on steroids…Rather than going to the drawing board to come up with a solution to a problem, the Rapid Equipping Force jumpstarts the process by evaluating what's already available commercially or in the production pipeline."[14] This process is advertised to accelerate the process of acquisition from "the traditional acquisition process that can take years rather than weeks or months."[15]

On the surface, this process may seem to meet the demand or fill the gap until contracted equipment is delivered, but commercially purchased equipment actually creates a vulnerability to the end user. What is available to United States forces commercially is also available to the enemy. These systems can be replicated or countered by any oppositional force. In addition, these COTS products can be purchased at reasonable costs, which make the short term solution a potentially devastating

---

[13] Donna Miles, "Rapid Equipping Force Speeds New Technology to Front Lines," *Defense AT&L*, November-December 2005, 47.

[14] Ibid, 48.

[15] Ibid, 48.

vulnerability. This is supported in Joint Vision 2020 which states, "We will not necessarily sustain a wide technological advantage over our adversaries in all areas. Increased availability of commercial satellites, digital communications, and the public internet all give adversaries new capabilities at a relatively low cost. [16]

## Interoperability

The disregard for the lack of interoperability between equipment is disconcerting. All of Armed Services realize this is a problem, but fail to recognize the significance of the issue. Even with guidance and direction from the Chairman of the Joint Chiefs of Staff (CJCS) through the Joint Requirements Oversight Council (JROC), units continue to disregard the necessity to ensure interoperability exists. In their defense, these units do everything possible to provide the warfighters with the capabilities; unfortunately, the procurement programs tend to acquire equipment only to meet the immediate demand. This process pays little or no concern to the implications of future combined operations.

Every Service is in the hunt to rapidly find and implement the latest technology that will add an additional capability to an existing system, increase the speed and accuracy of information output or offer a more streamlined information sharing system. However, many of these materials are being produced by different companies using different hardware that requires different power requirements and maybe even different operating systems. This creates a cascading effect as the demand for technological innovations continues to increase and vendors flood all levels of operations with new and improved capabilities. As Services purchase these materials for their individual needs through various acquisition processes, a large range of equipment is being introduced to

---

[16] Joint Vision 2020, 4.

the operational environment with little regard for compatibility. For example, the costly FBCB2 navigational equipment developed for the Army is "not interoperable with other systems such as the system used by the Marine Corps, the Mounted Digital Automated Communications Terminal (MDACT)."[17] Even as early as "late 2000, some $36 billion in planned acquisition reportedly would not be interoperable."[18] Despite realizing the need for interoperability between systems decades ago, the situation has not improved and it does not appear that the situation will improve in the near future. The QDR states, "The conventional acquisition process is too long and too cumbersome to fit the needs of the many systems that require continuous changes and upgrades–a challenge that will become only more pressing over time."[19] As joint operations become prevalent, this fact is even more important today to maintain interoperability between critical areas such as communications, intelligence and navigation.

Currently, a large portion of the operational technologies employed on the battlefield today are not only incompatible within the United States Armed Forces, but with allied and coalition partners. "Integration with Allies, particularly electronically, is becoming increasing complex and fraught, but nonetheless essential for those nations aspiring to fight in US-led coalitions. In fact, during the Kosovo operation in 1999, the different levels of electronic sophistication precluded seamless interoperability."[20] Unfortunately, as in this case, interoperability of equipment is often discovered during an

---

[17] Bennie Sanchez., *Fratricide, Technology and Joint Doctrine*, (Newport, RI, Naval War College, 2004), 7.

[18] Lisa Troshinsky, "DoD Creates Interoperability Coordination Program," *Navy New & Undersea Technology* (2 January 2001): 35.

[19] Ibid, xiv.

[20] R.K.Ackerman, "Kosovo Maps the Future of Information Technologies," *Signal* (December 1999): 49.

actual operation.

At this point, the lack of compatible systems now becomes a critical vulnerability that affects everyone on the battlefield. Bennie Sanchez recommends that "technological solutions need to focus on interoperability between services and connectivity among all units/players, including our coalition partners, if a Common Operational Picture is ever to be a reality."[21] David Price states,

> Interoperability problems go beyond just the difficulty of one system working with another. The different generations, modifications and upgrades multiply the number of equipment configurations end users are required to maintain and operate with limited manpower and support infrastructure. This further strains our training programs and the most obvious predetermined element, a limited budget."[22]

In this situation, versions of redundant capabilities must be monitored. Without a control mechanism to ensure systems can function together, the goal of interoperability will never be met. If corrections are not made immediately, units will continue to purchase for individual needs, the interoperability gap will continue to expand, and the potential for devastating effects will exist as more elements are required to work together on the battlefield. A conscious effort must exist when designing, developing and purchasing equipment to research existing products, collaborate with other Services and coalition partners, and to combine efforts across all Service spectrums when filling capability gaps.

It will take time to solve the insufficiencies of the current acquisition process. The elevated costs of technical equipment, the production process and allocation methods are all elements that a commander may not be able to change. However, commanders

---

[21] Sanchez, 13.

[22] Price, 10.

can make demands that will ultimately reduce the negative implications of an over-reliance on technology. Leaders must demand that only mature technologies are employed, prioritize technical capability requests that are absolutely required, pursue capabilities that can realistically be produced, and force military Services to ensure interoperability between equipment. If these steps are taken, the warfighters may not receive every capability demanded, but the products they receive will function, be interoperable and provide a capability that will be effective in the field.

# CHAPTER 5: LEADERSHIP

Military leaders today face a myriad of challenges such as increased training requirements, expanded scopes of responsibility and diverse working environments'; all while managing limited resources, budget cuts and manpower reductions. To assist with these challenges, commanders introduce advanced technology as a force multiplier to positively satisfy these leadership challenges. This chapter will focus on the necessity for a commander to effectively manage technical capabilities, the reduction in personal interaction and the declining leadership opportunities.

## Managing Technology

A challenge for leaders, not often achieved, is to balance mission requirements with available resources by managing the technology available for employment. Many times, the effective implementation of this technology by an organization often depends on the personality of the commander. John Corozza states that the "personality and leadership traits of the operational commander will ultimately determine how effectively this technology is integrated into the operational art of warfare toward achieving military objectives."[1] Despite these personality differences, commanders must ensure they understand the capabilities and limitations of each innovation and determine which tool is applicable to address the specific concern. Not all situations demand the same capabilities, and determining the right allocation of resources may be difficult when the commander does not know, or understand, the full spectrum of capabilities technical tools can provide.

The pressure to achieve mission accomplishment in conjunction with an existing

---

[1] Carozza, 18.

over-reliance on technology creates risks associated with misapplied innovation.

Unfortunately, the absence of this technical understanding can impede progress by adding

unnecessary fog and friction. William Holland supports this notion as he states, "no

commander will reap the benefits of modern equipment unless he knows how it works.

Technology makes a contribution only when operated by people who know how to apply

it."[2]

A mismanaged technical capability, not exempt from presenting critical

vulnerabilities, are elaborate information network systems that provide a vast amount of

intelligence to commanders. R. D. Williams explains the vulnerabilities associated with

this relationship:

> Experience, mistrust in the system and intuitive thought may provide a radically
> different interpretation from that achieved by the cold, rigid logic of
> microprocessor. Indeed, unless the system can filter effectively and with the trust
> of the commander, he may be subject to information overload and the processed
> data could act as a brake rather than speeding up his decision-making cycle.[3]

As these commanders attempt to manage the vast and overwhelming amounts of

information these systems contain, "there is a negative view, which sees HQs potentially

blundering through the decision making-process and taking longer to make poorer

decisions."[4] With such an overload of information it is difficult for a commander to

filter the relevant information from unnecessary details, which can essentially paralyze a

commanders' ability to make timely decisions. Access to unlimited information in

combination with a leaders' ability to make decisions does not translate to improved

---

[2] Holland Jr., 2.

[3] Williams, 49.

[4] J.Storr, "The Impact of Technology on War in the 21st Century," *RUSI International Security Review* (2000): 23.

efficiency. This is clearly supported in Joint Vision 2020 as it states, "Decision superiority does not automatically result from information superiority."[5]

## Personal Interaction

Regardless of the advantages current innovations have brought to the warfighter, they have concurrently persuaded leaders to reduce personal interaction with their subordinates. This is contrary to basic leadership principles that proved so effective during World War II. "General Patton's leadership in World War II offers a good example of the success that can result from personal interactions and observations. General Patton insisted on seeing the battle from the front lines rather than through the eyes of his staff."[6] In addition, "His physical presence on the battlefield in Europe allowed him to develop a strong unity of command due to personal relationships with his immediate subordinates."[7] "Through his physical contact with his subordinates, Patton was able to effectively convey his intent and accurately evaluate performance."[8]

General Patton's desire to be on the battlefield, to understand the conditions of the men, to know where his forces were through personal contact and to be able to make decisions based on information gained from personal experience all contribute to the product of good, basic leadership principles. These principles guided General Patton's efforts, allowing him to gain a realistic battlefield perspective and provided a situational awareness that could never be possible by any other means. The situational awareness he gained from this experience assisted in making effective and decisive decisions regarding

---

[5] Joint Vision 2020, 8.

[6] H. Essame, *Patton: A Study in Command* (New York, NY: Charles Scribner's Sons, 1974), 255.

[7] Carlo D'Este, *Patton: A Genius for War* (New York, NY: HarperCollins Publishers, 1995), 575.

[8] James A. McPherson, 1.

the enemy and current friendly situations.

Current technological innovations allow leaders to remain connected through electronic means from anywhere in the world. By taking advantage of these capabilities to lead from afar, it may appear that commanders are saving time, making effective decisions and improving efficiency; however, they are essentially sacrificing subordinate loyalty and confidence, face-to-face interaction and realistic situational awareness which go against basic leadership principles. R.L. Taylor states,

> "With emails that must be answered and PowerPoint presentations that need animating, the leadership within the United States Marine Corps does not spend as much time, as their pre-computer counterparts, getting to know those they may someday lead in combat. The distractions inherent in the form of technological advancement decrease opportunities to interact with one's Marines."[9]

Currently, it is typical for members of the same unit, located in the same building, and often in the same room, to have no personal contact or interaction during a normal duty day. This is made possible through the convenience of electronic and communications capabilities created through elaborate connectivity networks. This lack of interaction completely removes the personal touch and does not facilitate the necessity for interaction that has made previous fighting forces so effective. The camaraderie, esprit de corps, loyalty and the willingness to sacrifice life or limb for a teammate has slowly deteriorated, which can in some cases, can be contributed to a technological over-reliance.

## Limited Leadership Opportunities

When technology decreases or impedes individual leadership opportunities, a negative effect on personnel capabilities becomes a threat. This is supported by the

---

[9]Taylor, 3.

amount of computer based training (CBT) the armed forces are pursuing. Marines refer

to this transition as

> Another leadership opportunity lost tied directly to the ever increasing reliance on computer-based training. What was once a Marine's job, to stand in front of a class and instruct, is being outsourced to a computer. Fewer, are the formal instruction forums which could have benefitted the instructor, becoming a subject matter expert, and public speaking exposure. [10]

Every Service has an individually unique system for providing CBT

opportunities. The Navy and Army have Navy/Army Knowledge Online (NKO) that

provide an outlet to conduct job specific, security clearance, and even college-level

course training. And the Services are increasingly relying on this method of instruction

as a primary means to reduce costs and increase qualifications by reaching more Service

members. Contrary to the effort, reduced personal interaction and decreased leadership

opportunities are a result of these technical education avenues.

Although technology has allowed commanders to limit personal interaction and

decrease leadership opportunities, it does not mean leaders must continue down this path

of over-reliance. As discussed above, these capabilities left uncontrolled lead towards

negative effects that lure commanders into reducing the strength of personal leadership.

As R.L. Taylor explains, the "current over-reliance on technological capabilities limits

leadership opportunities, promotes a divisive atmosphere and at times has hindered

combat operations. A degeneration of leadership and teamwork as well as reputation is

bound to ensue if this trend toward techno-adoration goes unchecked." [11] Based on this

concern, commanders must make every effort to aggressively educate themselves on

effective technological management and implementation, gain personal situational

---

[10] Taylor, 4.

[11] Taylor, 2.

awareness through subordinate interaction and maximize leadership opportunities for subordinates. Most importantly, leaders must realize that despite the advantages provided by technology, may provide, nothing can substitute for good leadership and human interaction.

## CHAPTER 6: PERSONNEL

In the area of personnel and manning, there is a risk required operator skill levels will exceed the aptitude of current Service members and new recruits. This becomes a vulnerable situation when Service members are faced with the requirement to operate advanced technical systems that are beyond their skill levels. This chapter will discuss the technical aptitude required of current and future Service members and the vulnerabilities associated with the over-reliance on technology.

### Technical Aptitude

Technology is continuously becoming more advanced with extremely sophisticated systems that require higher skill levels to operate. And in today's military it is almost a requirement for all Service members to have some level of technical operating capability. J.D. Fletcher agrees by saying, "The complexity of military operations has continued to increase along with the human performance needed to operate, maintain, and deploy the material, devices, and equipment they employ."[1] Unfortunately, not every Service member is suited for specific operating systems. In an article discussing learning and technology, Malcolm Brown says that "reliance on technology returned to the old transmission model of 'one size fits all.' Essentially, not all students are adept with technology or new media."[2] Despite this disparity, the armed forces continue to incorporate these systems into every aspect of daily operations regardless of the aptitude of Service members requiring technical skills.

---

[1] J.D. Fletcher, *Virtual Reality: Training's Future?* Edited by Seidel and Chatelier (Plenum Press, New York, 1997), 169.

[2] Malcolm Brown, "Learning and Technology–"In That Order," *EDUCAUSE Review,* vol. 44, no. 4 (July/August 2009): 62.

Not only does the military take a risk by employing technically under-qualified individuals, but organizations are now forced to consolidate multiple jobs into one position due to current fiscal constraints, adding to increased demands for higher levels of individual and organizational performance. In an article discussing the sophistication of naval warships Jean Grace explains, "Ships also will have smaller crews so every sailor will be trained for multiple jobs,"[3] which places additional technical responsibilities on personnel that are already pressed to the limit, and often beyond their personal skill level. And this does not apply to only junior Service members or new recruits. This affects personnel at every level in the chain of command. To support this, Raymond Millen states that "even experienced commanders do not have the acquired skills,"[4] when dealing with sophisticated technical equipment. And, "if managed ineffectively by the operational commander, technology can lead to reductions in the efficiency and effectiveness of military forces employment."[5] Bottom line, "Human supervisors must know the logic and the limits of these machines if they are to use them as operational artisans."[6] Unfortunately, as this trend continues, "the required expertise goes up, not down, as the environment becomes more technical."[7]

If the personnel leading the armed forces do not have the aptitude for utilizing technology to its fullest extent, dangers exist that may effect all military operations. Extensive technical education programs should be instituted immediately along with advanced recruiting methods. These methods should include specialized testing for new

---

[3]Grace Jean, "Hybrid Sailors," *National Defense* (May 2007): 34.

[4] Millen, 36.

[5] Carozza, 17.

[6] Holland, 2.

[7] Ibid, 2.

recruits in order to measure their technical aptitude and targeted recruiting for those youths proficient in technical areas needed in the military.

# CHAPTER 7: COUNTERMEASURES

What has been briefly mentioned in many of the previous chapters are the countermeasures associated with the dependence on technology. Regardless of the technological domination American armed forces have over their adversaries, the enemy will always "get a vote." Despite the United States strategic guidance to develop technology as a means to support and defend the nation's interests, the enemy is persistently preparing to counter these advanced capabilities. Therefore, the enemy's will to counter US technical capabilities exists in almost every aspect of the doctrine, organization, training, materiel, leadership, personnel and facilities construct.

In an effort to deny the enemy the opportunity to counter existing technology, precautionary devices such as back-up power supplies, encryption codes and firewalls are often incorporated into the equipment. Unfortunately, regardless of how elaborate these anti-countermeasure systems may be, they still remain vulnerable to attack. More alarming is the realization that even America's most secure networks are vulnerable to operational failure. Arnaud de Borchgrave supports this when he states, "Secure networks are vulnerable. National Security Agency (NSA) personnel have identified theoretical vulnerabilities of the Secret Internet Protocol Router Network (SIPRnet) that carries much of the military's command and control communications via the Global Command and Control System (GCCS), as well as other sensitive information." [1] This is alarming considering that the majority of US classified information sharing and coordination is conducted on the SIPRnet. If this network is vulnerable, how is the American military supposed to keep its systems secure?

---

[1] Arnaud de Borchgrave et al., *Cyber Threats and Information Security: Meeting the 21st Century Challenge,* (Washington, Center for Strategic and International Studies, December 2000), 48.

In an additional supporting argument Vermon Loeb states in an article in the Washington Post Magazine, "The Department of Defense information technology (DODIT) infrastructure, including its intelligence infrastructure, regularly fails during normal operations without help from enemies. A particularly serious failure virtually shut down the headquarters of the National Security Agency (NSA) at Fort Meade, for over three days."[2] If these systems are susceptible to failure outside the actions of adversaries, imagine the devastation that could be accomplished in a coordinated attack.

In response to this suggestion, Robert Scales says, "A thinking opponent will quickly realize that our intensive reliance on information age technologies becomes a weakness that can become an asymmetric target."[3] The enemy realizes American forces are the most technologically advanced military in the world. This dependence provides an opportunity for adversaries to develop and implement alternative measures to erase their overwhelming disadvantage. The enemy understands that "Even where the US has technical advantages, effective countermeasures usually exist. In some cases these degrade US capabilities directly. In other instances, adversaries operate in politico-military arenas beyond the scope of US military capabilities, rendering the technology irrelevant."[4] So despite maintaining battlefield technological dominance, it does not prevent the enemy from conducting successful attacks against American forces with either technical countermeasures or through indirect means.

In its Global Trends 2015 estimate, the National Intelligence Council assessed "Adversaries will seek to attack US military capabilities through electronic warfare,

---

[2] Vermon Loeb, "Test of Strength," *The Washington Post Magazine*, 29 July 2001, 9.

[3] Robert H. Scales, Jr, *Future of Warfare*, (Pennsylvania, U.S. Army War College, 1999), 120.

[4] Gentry, 89.

psychological operations, denial and deception, and the use of new technologies such as directed energy weapons or electromagnetic pulse."[5]  These adversaries are currently executing thousands of coordinated network attacks around the country daily, and these countermeasures are not always complex or expensive.  They have used simple, inexpensive devices to defeat the most advance technology American armed forces bring to the battlefield.  These devices include methods to illuminate the darkness that directly disable the advantages of fourth generation night vision equipment.  They install basic road blocks to defeat high-tech navigational equipment that does not recognize obstacles.  They defeat the most advanced electronic warfare (EW) capabilities by incorporating remote detonating wires to roadside explosives.  All of these alternative targeting techniques are direct counters to America's over-reliance on technology.  The enemy knows US armed forces have an over-reliance on technology and will use the dependence as an outlet to disrupt or defeat America's technical dominance.

A more specific US capability that is extremely susceptible to enemy attack is the global positioning system (GPS) network.  As discussed in previous chapters, military forces depend on some form of GPS capabilities for navigation, precision weapons and tracking elements in the field.  Currently, the over-reliance and dependence American armed forces place on this system makes it one of the most technologically vulnerable capabilities.  Unfortunately, the enemy often has the capability and access to equipment which can severely degrade this system.  According to the Weapon Systems Technology Information Analysis Center,

---

[5] National Intelligence Council. *Global Trends 2015: A Dialogue About the Future With Nongovernment Experts* (December 2000), 57.

There seems to be a broad consensus among the experts, that wide-band noise jamming represents the most affordable, tactically feasible, effective jamming technique that is likely to be encountered by GPS receivers in the near term. With numerous sources of information available (many on the internet), a competent electronics technician should be easily able to build a noise jammer with an effective range of tens of kilometers for $1K to 10K.[6]

The analysis continues by stating that "These jammers could easily degrade precision munition accuracies from the 10m or better accuracy associated with GPS guidance, out to the tens to hundreds of meters associated with uncorrected internal guidance (depending on weapon flight duration)."[7] This vulnerability presents the potential for devastating results beyond risk tolerance levels.

American armed forces must keep in mind that these attacks are not only limited to state actors with a formal military organization and thriving economies. As more non-state actors and violent extremist organizations (VEOs) improve funding avenues and increase global support, it becomes more realistic for these units to employ these countermeasures. Unfortunately, as technical capabilities advance, technology becomes "a double-edge sword. On the one hand, technological and social changes are making war more costly for modern democracies. But at the same time, technology is putting new means of destruction into the hands of extremist groups and individuals."[8] And as these adversaries develop the ability and knowledge to produce countermeasures, American dominance in the technological field will eventually be reduced.

---

[6] Mark Scott, "Anti-Jam GPS Pat III: Protecting Weapon Receivers from Jamming," *Weapon System Technology Information Analysis Center Quarterly,* Volume 3, Number 3 (July 2002): 4.

[7] Ibid, 7.

[8] Joseph S. Nye, Jr, *Soft Power: The Means To Success In World Politics*, (New York, Public Affairs, 2004), 21.

# CHAPTER 8:  RECOMMENDATIONS

This chapter will offers recommendations on how to reduce the vulnerabilities and risks associated with the current over-reliance on technology in the American military.  The format will remain true to the doctrine, organization, training, materiel, leadership, personnel and facilities (DOTMLPF) convention utilized in the previous argument.

## Doctrine

It is important for leaders to understand the dynamic relationship between technology and doctrine.  Throughout the discussion, it was identified that each has the flexibility to remain independent of each other, but yet should not change in complete isolation.  Unfortunately, as the over-reliance on advanced technology becomes a significant influence in military operations, technology often outpaces doctrine and becomes the driving factor on how daily activities are conducted.  However, this is not necessarily a negative situation.  But through a proper understanding of this relationship and sufficient control measures, the possibilities for destructive implications can be avoided.

To prevent this from happening, it is necessary for leaders to understand that doctrine should generally drive how operations are conducted, and that advanced technology is merely a tool to enhance doctrinal implementation.  In the event technology actually does outpace doctrine, doctrine must be adjusted and brought up to pace accordingly in order to provide effective guidance for the operational procedures of the new technical capability.  Without this adjustment, vulnerabilities exist that create dangers and hazards that often end with devastating results.

In addition, leaders must avoid the temptation to implement technology as the quick and easy solution in the event doctrinal procedures fall behind. This premature reaction will result in technical capabilities that have no operational guidance. Leaders must understand that technology is not the answer for every situation and without baseline operational procedures technology is essentially useless.

Acquisition programs are in place to ensure technologies are effective, and there are organizations which keep doctrinal procedures current. However, more careful collaboration between these two organizations needs to exist in order to ensure the two maintain a close working relationship. With such a rapid influx of advanced technology into military operations, without a control measures, doctrine is likely to be neglected.

**Organization**

As technology leads to new and improved methods for increasing efficiency, reducing manpower strains and obtaining a marked advantage over the enemy, an over-reliance on these capabilities forces organizational changes that negatively effect operations at all levels. Of these changes, advanced technology has provided capabilities that give leaders the option to establish a centralized command and control (C2) and to increase an already over-extended span of control.

It is essential for leaders to resist the temptation to create a centralized C2 based on the availability of advanced technological capabilities. Commanders must resist the temptation to make decisions based on the unlimited access to unfiltered information. This, in turn, creates a micro-management situation that is not beneficial for anyone in the chain of command.

In addition, technology provides commanders the capability to stay connected with subordinates throughout their area of responsibility with a touch of a button. This situation leads to the creation of a centralized chain of command, which in today's military, creates multiple issues that negatively effect an organization. Most importantly, this vertical centralized command structure allows for maximum flow of unfiltered information to the commander that often creates a false situational awareness. It is important to commanders to understand that unlimited access to information and the ability to communicate with subordinates through advanced communications systems does not qualify a commander to make every decision.

As commanders are faced with large geographical responsibilities, they must create solutions to manage the abundant operational requirements. Unfortunately, as these requirements continue to increase, commanders are forced to spread their forces throughout their areas of responsibility. To assist with the C2 of these forces, commanders rely on technologies such as the video teleconference (VTC) and satellite communications as the solution. Unfortunately, leaders often get caught up in the false security that these systems will function in every circumstance and can be utilized confidently as the link between all forces. This confidence then gives commanders the feeling that they can continue to extend their spans of control and facilitate engagements throughout larger areas in order to meet theater endstates.

To resist this temptation, commanders must weigh the benefits against the risks before spreading forces beyond operational reach, especially to such an extent that all forces cannot be reached in sufficient time. Just because a capability exists, does not mean that it has to be used. And especially in the case of an increased span of control, extreme consequences exist if the technology fails.

## Training

A commander must promote a training environment with a focus on achieving a basic foundation of skills before introducing advanced technology into the training regiment. Leaders must ensure that each and every service member thoroughly understands every intricate detail of the equipment and progressively implements technology as core skills are acquired.

An elevated operational tempo (OPTEMPO) should not be used as a crutch to excuse the requirement to train to basic operational skills. Regardless of compressed training timelines, leaders must focus on the basic standards that must be met before introducing technology. Although incorporating technology early during training provides significant advantages to a commander, the priority should be to teach every operator basic individual survival skills before moving to advanced capabilities. This will provide each Service member with a baseline operational skill to fall back on in the event technology fails.

Leaders must also incorporate as much live training as possible. A computer generated system, or simulator, simply cannot duplicate the physical aspect, the physical exertion required or the tension and stresses that realistic training provides. Live training provides the most realistic environment while maintaining safety measures to ensure each operator can function in a high stress, life threatening situation. The skills obtained during live-fire evolutions can never be fully simulated by an automated system.

Once technology is actually incorporated into training, Service members must be capable of compensating during a situation when technology fails or is inoperable. They must possess the inherent skills required to continue operations without the dependency on technology, and how to execute trouble shooting procedures without hesitation. Sufficient time must be allocated towards ensuring each technical operator knows their equipment inside and out, and what to do if they cannot fix the problem. Repeated drills must be run, trouble shooting exercises must be held, and members should be tested on their ability to operate without technical enhancements. Without these skills, the risk is too great and the potential consequence of technological failure is too dire.

### Materiel

The rising costs of sophisticated technological capabilities play a major role in the production of the most advanced technology required by the warfighters today. Many of these programs exceed acceptable funding limits and are often produced in such limited quantities that production never meets the demand on the battlefield. It is important to focus the priority of innovation developments towards effective and lucrative programs that do not end up over budget or cancelled. To prevent this, controls need to be placed on the reduction of outlandish and extravagant programs that aim to meet the demands of current and future threats. This does not suggest that all high-cost programs be cancelled but all programs, especially expensive ones with elevated expectations of success, should be scrutinized to ensure effectiveness and cost efficiency.

Every effort must be used to avert the stovepipe production phenomenon. When possible, attempt to consolidate production within one vendor or at least minimize the number of players involved in the production of a single system. This will improve accountability, allow for sufficient testing under one vendor and minimize deficiencies when all the pieces are put together. However, it is not essential to production to completely eliminate stovepipe production. But when possible, it should be prevented.

In addition, when stovepipe production exists, it is essential that coordinated oversight be in place to provide quality assurance measures that may prevent unnecessary problems or delays.

A threat to the warfighters, despite showing great initiative, is the deployment of immature technologies. Regardless of the demand and the requirement of technology on the battlefield, it is unacceptable to provide operational equipment that has not been properly tested before implementation. This initiative does not compensate for the potential for technology to fail or perform inadequately when the warfighters need it most. And it definitely does not consider the potential catastrophic risk weighed against the consequence of failure when the reliability of a capability has not been properly field tested. To prevent this, quality assurance teams must be required to mandate proper testing and demand the test results prove sufficient by military standards before deploying these technologies.

Interoperability between operational equipment is a critical requirement on today's battlefield, but is too often ignored when considering system compatibility between individual units, Services and coalition partners. Unfortunately, the existing Chairman Joint Chiefs of Staff (CJCS) requirement to ensure compatibility is completely overlooked in most cases. Many acquisition programs are only concerned about individual performance and do not consider the current battlefield requirements to operate with multiple organizations. The Joint Capabilities Integration Development System (JCIDS) exists to prevent this from happening; however, more emphasis needs to be placed on interoperability. This will be a difficult task considering the large number of vendors producing technical equipment that varies only slightly, but does not operate with other systems. These vendors are only interested in sales, and not ensuring interoperability is achieved. Extreme emphasis should be directed towards a solution to this problem.

## Leadership

A basic principle of leadership is to know your people. But with today's innovations such as email, PowerPoint briefings, video teleconferences, and satellite communications, commanders can easily separate themselves from their subordinates. It is crucial to avoid this segregation as much as possible and make every effort to have face-to-face time with the members of the command.

Commanders must resist the temptation to utilize communications technology as a means to make front-line decisions. Information or intelligence received through electronic means does not equate to a true operational picture, and commanders should not treat it as such. Commanders must place trust and confidence in their subordinates to make accurate and informed decisions based on their physical relationship to the situation on the ground.

Commanders must learn how to properly manage the technology available to their organizations, which will result in proper employment. Understanding doctrine and knowing the capabilities of these technologies will assist in using them to meet the military objectives and endstates vice utilizing technology to a disadvantage. If the proper use of the technologies is uncertain, refrain from using the capability until a complete understanding is achieved. In this case, a commander must leverage and include the knowledge of his subordinates to assist in technological implementation decisions.

Every commander must maximize personal leadership opportunities for his subordinates despite the movement towards tools such as computer based training (CBT). CBT reduces the opportunities for leaders to become comfortable in a teaching environment, removes the personal touch associated with instruction and limits the interaction with peers. In many cases, CBT offers a unique and efficient method, at the expense of being effective, towards reaching a large number of individuals for qualifications and to maintain a baseline operational proficiency. However, commanders should seize every opportunity to provide an environment that promotes close teacher-student interaction. This interaction better prepares leaders for future advancement, teaching experience and maintains a leadership foundation for years to come.

## Personnel

As the requirement for a basic technical knowledge increases, so does the requirement for personnel to have the aptitude for technology. Not every individual is suited for operating elaborate and sophisticated technological equipment. But as the military becomes more reliant on technology, it is essential to have the right people in the right position. In order to ensure this, more specific and targeted recruitment and screening should be included to determine the technical aptitude of an individual.

In addition, Service members currently serving a military obligation should not be exempt from this testing. New technical capabilities may have been incorporated into particular employment specialties that did not previously require a technical aptitude. Individuals in these specialties should also be evaluated to determine if they meet the qualifications to operate the required new advanced technology. If they don't, training to the appropriate level of proficiency should be mandatory.

An effort should be made to minimize the sophistication and intricacy of new technical capabilities. The proficiency of operators will not be equal across all levels of the military; therefore, minimizing the necessity to have advanced trained Service members will ease the burden of requiring highly trained technical operators.

## CONCLUSION

From the strategic to the tactical level, in one form or another, technology has flooded the gates with capabilities warfighters cannot seem to operate without. Whether an operator is searching for the most effective method to find the way from one location to another or a commander is leading Service members from thousands of miles away, American troops are finding themselves deep in a technological age that has arguably taken over the ways military forces fight. Without proper control and effective measures to prevent over-reliance on technology from continuing, military forces will rapidly come to the realization, possibly in combat, that such a dependency will increasingly produce adverse battlefield effects across all spectrums of warfare.

To address this issue, the author utilized the DOTMLPF construct as a convention to conduct a broad analysis of the current over-reliance on technology. In each of the chapters, concerns with the implementation of technology have been identified and discussed in detail. Additional concerns were addressed by directly discussing potential vulnerabilities and countermeasures. This broad based approach was used to describe the prevalence of the current over-reliance on technology, and to effectively highlight many of the critical areas that create negative effects. To go into further detail in any of the areas covered would have been beyond the scope of this paper.

Leaders must understand that technology is here to stay, and will most likely become more influential over the next several decades. If the points addressed in this paper are considered, an over-reliance on technology may be assessed, controlled and monitored as new capabilities enter the military's inventory. It is up to the leadership to ensure technology both meet the demands of the warfighter and is effectively incorporated into current operating systems while recognizing the negative effects

innovation can create.

# BIBLIOGRAPHY

Ackerman, R.K. "Kosovo Maps the Future of Information Technologies." *Signal* (December 1999): 49.

Adams, Thomas K. "GPS Vulnerabilities." *Military Review, 81* (March-April 2001): 10-16.

Altmann, Jurgen. "Critical Analysis of New Weapons Technologies." *Peace Review: A Journal of Social Justice* (Date Unknown): 144-154.

Ayers III, William. "Fratricide: Can It Be Stopped?" *Globalsecurity.org.* November 23, 2003. http://www.globalsecurity.org/military/library/report/1993/AWH.htm (accessed February 24, 2011).

Bateman, Robert L. *War: A View From the Front Lines.* California, Presidio Press, 1999.

Beason, J. Douglas, and Dr. Lewis, Mark, "The War Fighter's Need for Science and Technology." *Air & Space Power Journal* (Winter 2005): 71-81.

Berg, Paul D. "Doctrine and Technology." *Air & Space Power Journal* (Spring 2008): 20.

Bolia, Robert S. "Overreliance on Technology: Yom Kippur Case Study." *Parameters* (Summer 2004): 46-56.

Bonsignore, Ezio. "A Technology Too Far: "Reviewing" (or Terminating?) the FCS Programme." *Military Technology* (July 2009): 12-14.

Borchgrave, Arnaud. *Cyber Threats and Information Security: Meeting the 21st Century Challenge.* Washington, Center for Strategic and International Studies, December 2000.

Brown, Malcolm. "Learning and Technology –"In That Order." *EDUCAUSE Review Magazine*, Vol. 44, No. 4 (July/August 2009): 62-63.

Canby, Steven L. *New Conventional Force Technology and the NATO-Warsaw Pact Balance.* New Technology and Western Security Policy. Basingstoke, UK: Macmillan, 1985.

Carozza, John L. *The Unspoken Consequence of Command, Control, Communications Technology: Enhanced Micromanagement by Risk-Averse Commanders.* Final Paper. Newport, R.I.: Naval War College, 9 February, 2004.

Cebrowski, Arthur K. and Garstka, John J. "Network-Centric Warfare: Its Origin and Future." *Proceedings* (January 1998): 28-35.

Chairman of the Joint Chiefs of Staff (CJCS). *Joint Vision 2010.* Washington, D.C. Government Printing Office: July 1996.

--- *Joint Vision 2020.* Washington, D.C. Government Printing Office: June 2000.

Clausewitz. *On War.* Princeton, NJ: Princeton University Press, 1986.

Clubb, Glen E. "The Sensor Irony: How Reliance on Sensor Technology is Limiting Our View of the Battlefield." Master's thesis, Norfolk, VA: Joint Forces Staff College, Joint Advanced Warfighting School, 2010.

Dale, Jon. *"GPS Capabilities for the Warfighter."* Newport, RI: Naval War College, February 12, 1996.

Davis, Mark G. "Operation Anaconda: Command and Confusion in Joint Warfare." Master's thesis, School of Advanced Air and Space Studies, Air University, Maxwell Air Force Base, Alabama, 2004.

D'Este, Carlo. *Patton: A Study in Command.* New York, NY: Harper Collins Publishers, 1995.

Dizard, John. "GEKKO." *National Review* (August 11, 1997): 28-29.

Dumond, John. *Maturing Weapon Systems for Improved Availability at Lower Costs.* Santa Monica, CA: RAND (1994).

Dunlap, Charles J. "Special Operations Forces After Kosovo." *Joint Force Quarterly, 28* (Spring/Summer 2001): 7-12.

--- *Technology and the 21st Century Battlefield: Recomplicating Moral Life for the Statesman and the Soldier.* Carlisle Barracks, PA: U.S. Army War College, January 15, 1999.

Erwin, Sandra L. "Ground Connections: Mobile Broadband for Roaming Troops: Pipe Dream or Reality?." *National Defense* (June 2008): 48-50.

--- *"Threat to Satellite Signals Fuels Demand for Anti-Jamming Products."* Washington, DC: National Defense University, June 2000.

Essame, Hubert. *Patton: A Study in Command.* New York, NY: Charles Scribner's Sons, 1974.

Farlex. "The Free Dictionary." http://www.thefreedictionary.com/organization (accessed March 9, 2011).

Featherston, Donald.F. The Bowmen *of England: The Story of the English Longbow.* Barnsley, UK: Pen & Sword Press, 2003. Fletcher, J.D. *Virtual Reality: Training's Future?* Edited by Seidel and Chatelier. Plenum Press, New York, 1997.

Fletcher, J.D. *Virtual Reality: Training's Future?* Plenum Press, New York, 1997.

Gentry, John A. "Doomed to Fail: America's Blind Faith in Military Technology." *Parameters* (Winter 2002-2003): 88-103.

Hallion, Richard D. "Doctrine, Technology, and Air Warfare: A Late Twentieth–Century Perspective." *Airpower.au.af.mil.* Fall 1987. http://www.airpower.au.af.mil/AIRCHRONICLES/apj/apj87/fal87/hallion.html (accessed February 24, 2011).

Hammes, Thomas X. *The Sling and the Stone: On War in the 21st Century.* St. Paul, MN: Zenith Press, 2006.

Herbert, Adam J. "Army Change, Air Force Change." Air *Force Magazine, 89* (March 2006): 36-41.

Hoey, David and Benshoof, Paul. *Civil GPS Systems and Potential Vulnerabilities.* Technical Report. Eglin Air Force Base, FL: Air Armament Center, October 25, 2005.

Holland Jr., William J. "Technology Is Key to the Operational Art, Not an Obstacle." *Proceedings* (April 1, 2004): 2. http://www.proquest.com/ (accessed February 17, 2011).

Iannotta, Ben. "Agencies try to calm fears about GPS problems." *Airforcetimes.com.* September 13, 2009. http://www.airforcetimes.com/news/2009/09/airforce_gpsupdate_092109w/ (accessed February 24, 2011).

Ignatieff, Michael. *Virtual War: Kosovo and Beyond.* New York: Metropolitan Books, 2000.

Jean, Grace. "Hybrid Sailors." *National Defense* (May 2007): 34-36.

Storr, J. "The Impact of Technology on War in the 21st Century." *RUSI International Security Review* (2000).

Kaufman, Randy L. *Precision Guided Weapons: Panacea or Pitfall for the Joint Task Force Commander?* Final Paper. Newport, R.I.: Naval War College, February 3, 2003.

Komarow, Steven. 1999. "Army Forces to See Major Restructuring." *USA Today*, February 16, 1999, A1.

Lambeth, Benjamin S. "Desert Storm and Its Meaning: The View From Moscow." *RAND* (October 5, 1992).

Loeb, Vermon. "Test of Strength." *The Washington Post Magazine* (July 29, 2001): 9-10.

Martin, Zachary D. "Tempo, Technology, and Hubris: The Garrison Mentality in Iraq." *Marine Corps Gazette* (May 2007): 50-55.

McClure, William B. *Technology and Command: Implications for Military Operations in the Twenty-first Century.* Occasional paper. Maxwell Air Force Base, AL: Center for Strategy and Technology, Air War College, July 2000.

McPherson, James A. *Operation Anaconda: Command and Control through VTC.* Newport, RI: Naval War College, February 12, 2005.

McPherson, Michael R. *GPS and the Joint Force Commander: Critical Asset, Critical Vulnerability.* Newport, RI: Naval War College, May 18, 2001.

Metz, Steven. "Strategic Asymmetry." *Military Review*, 81 (July-August 2001): 23-31.

Miles, Donna. "Rapid Equipping Force Speeds New Technology to Front Lines." *Defense AT&L* (November-December 2005): 47-49.

Millen, Raymond. "The Art of Land Navigation GPS Has Not Made Planning Obsolete." *Infantry Magazine* (January-April 2000): 36-43.

Moorehead, Richard D. "Technology and the American Civil War." *Military Review* (May-June 2004): 61-63.

National Intelligence Council. *Global Trends 2015: A Dialogue About the Future With Nongovernment Experts*, December 2000, 57.

Nye, Joseph S. Jr. "*Soft* Power: The Means To Success In World Politic." New York: Public Affairs, 2004.

Onley, Dawn S. "Systems help avert friendly fire deaths." *gcn.com.* May 16, 2003. http://gcn.com/Articles/2003/07/24/Systems-help-avert-friendly-fire-deaths.aspx (accessed February 24, 2011).

Paone, Chuck. "Chief of Staff Calls for Harmony between Technology, Doctrine." *Af.mil.* October 1, 2009. http://www.af.mil/news/story.asp?id=123170771 (accessed February 24, 2011).

Price, David R. *Technology In Transformation: Critical Strength or Critical Vulnerability.* Newport, RI: Naval War College, May 18, 2001.

Callum, Robert. "Will Our Forces Match the Threat?" *Proceedings* (August 1998): 50.

Sanchez, Bennie. *Fratricide, Technology and Joint Doctrine.* Final Paper. Newport, RI: Naval War College, February 9, 2004.

Sanderson, Jeffrey R. *"General George S. Patton, Jr: Master of Operational Battle Command. What Lasting Battle Command Lessons Can We Learn From Him?"* Monograph, School of Advanced Military Studies, United States Army Command and General Staff College, Fort Leavenworth, KS, 1997.

Scales, Robert H. *Future of Warfare.* Carlisle Barracks, PA: US Army War College, 1999.

--- *"Trust, Not Technology, Sustains Coalitions."* Carlisle Barracks, PA: US Army War College, December 1999.

Schnaubelt, Christopher M. "Whither the RMA?" *Parameters* (Autumn 2007): 95-107

Scott, Mark. "Anti-Jam GPS Pat III: Protecting Weapon Receivers from Jamming." *Weapon System Technology Information Analysis Center Quarterly,* Volume 3, Number 3 (July 2002): 4-8.

Simon, Herbert A. "Information 101: It's Not What You Know, It's How You Know It." *Journal for Quality and Participation* (July-August 1998): 30-33.

Skaggs, Michael D. "Digital command and control: Cyber leash or maneuver warfare facilitator?" *Marine Corps Gazette 87* (Jun 2003): 46.

Solivan, Austin C. "The Advanced Course Gets More Advanced." *Quantico.usmc.mil.* June 14, 2010. http://www.quantico.usmc.mil/Sentry/StoryView.aspx?SID=4220 (accessed February 24, 2011).

Storr, J. "The Impact of Technology on War in the 21st Century." *RUSI International Security Review* (2000): 23.

Talton, Trista. "Corps tackles erosion of navigation skills." *Marinecorpstimes.com.* June 7, 2009. http://www.marinecorpstimes.com/news/2009/06/marine_land_nav_gps_060709 w/ (accessed February 24, 2011).

Taylor, R.L. *Technology Sound not Technology Bound: The risks of over-reliance on modern military capabilities.* Quantico, VA: United States Marine Corps, Command and Staff College, February 19, 2009.

The Editors. "Technology and Logistics." *Air Force Journal of Logistics, Vol. XXX, Number 1,* (Spring 2006): 45.

The Editors. "US Army Battle Command Systems." *Military Technology* (October 2009): 48-54.

Troshinsky, Lisa. "DoD Creates Interoperability Coordination Program." *Navy News & Undersea Technology* (January 2, 2001): 35.

Tucker, David. "The RMA and the Interagency: Knowledge and Speed vs. Ignorance and Sloth?" *Parameters, 30* (Autumn 2000): 66.

Uhler, Dale G. "Technology Force Multiplier for Special Operations." *Military Technology, Special* Edition (2009): 42-48.

Unknown. *"Avoiding Over-Reliance on Technology."* Research Paper. Quantico, VA: Marine Corps University, 2004.

Unknown. "US Army Battle Command Systems: Current Status and Future-Oriented Modernisation Strategy." *MILTECH* (October 2009): 48.

U.S. Air Force. *Air Force Basic Doctrine.* Air Force Field Doctrine 1. Secretary of the Air Force. Washington, DC: November 17, 2003.

U.S. Department of the Army. *Operations.* Field Manual 100-5. Headquarters, Department of the Army. Washington, DC: June 14, 1993.

--- *Operational Terms and Graphics.* Field Manual 1-02. Headquarters, Department of the Army. Washington, DC: September 21, 2004.

U.S. Department of Defense. *Dictionary of Military and Associated Terms*, Joint Publication 1-02. Joint Chiefs of Staff. Washington, DC: April 12, 2001.

--- *National Defense Strategy.* Washington, DC, June 2008.

--- *Quadrennial Defense Review Report.* Washington DC: Government Printing Office, February 2010.

U.S. Joint Chiefs of Staff. *Joint Operation Planning.* Joint Publication 5-0. Joint Chiefs of Staff. Washington, DC: December 26, 2006.

U.S. Joint Forces Command. *"Kosovo After-Action Report to Congress, Executive Summary."* Joint Center for Lessons Learned Quarterly Bulletin, III (June 2001).

U.S. President. *National Security Strategy.* The White House. Washington, DC. May 2010.

U.S. Secretary of the Air Force. *Air Force Doctrine Document 1.* Secretary of the Air Force. Washington DC: November 17, 2003.

Vanderbeek, Walter A. *Technology or Doctrine, Search for Balance.* Core Course Paper. Fort Leslie, WA: National War College, 1994.

Van Riper, P.K. "Information Superiority." *Marine Corps Gazette* (June 1997): 54-62.

Vego, Milan. "Operational Command and Control in the Information Age." *Joint Forces Quarterly 35* (2004): 100-107.

--- *Operational Warfare.* Newport, RI: Naval War College, 2000.

--- "Technological Superiority is NOT a Panacea." *Proceedings 136, no. 10* (October 2010): 28-32.

Williams, D. "Is the West's Reliance on Technology the Panacea for Future Conflict or its Achilles Heel?" *Defense Studies, Vol.1. No.2* (Summer 2001): 38-56.

Wooldridge III, Tyler E. "Order a PowerPoint Stand-down." *Proceedings, Vol. 130, Issue 12* (December 2004): 85.

# GLOSSARY OF ACRONYMS

DOTMLPF:  Doctrine, Organization, Training, Materiel, Leadership, Personnel, Facilities
C2:  Command and Control
C3:  Command, Control and Communications
CBT:  Computer Based Training
CFLCC:  Combined Forces Land Component Commander
CFACC:  Combined Forces Air Component Commander
COA:  Coarse of Action
COTS:  Commercial-Off-The-Shelf
CJCS:  Chairman Joint Chiefs of Staff
DoD:  Department of Defense
EW:  Electronic Warfare
FBCB2:  Force Battle Command Brigade and Below
FCS:  Future Combat System
GCCS:  Global Command and Control System
GPS:  Global Positioning System
IT:  Information Technology
JCIDS:  Joint Capabilities Integration Development System
JROC:  Joint Requirements Oversight Council
JSOTF:  Joint Special Operations Task Force
K2:  Karshi Khanabad
MDACT:  Mounted Digital Automated Communications Terminal
NKO:  Navy Knowledge Online
NSA:  National Security Agency
ODS/DS:  OPERATION DESERT STORM/DESERT SHIELD
OEF:  OPERATION ENDURING FREEDOM
OIF:  OPERATION IRAQI FREEDOM
OPTEMPO:  Operational Tempo
PC:  Personal Computer
PGM:  Precision Guided Munition
QDR:  Quadrennial Defense Review
RDO:  Rapid Decisive Operations
RMA:  Revoluton of Military Affairs
SIPRNet:  Secret Internet Protocol Router Network
TACSAT:  Tactical Satellite
VEO:  Violent Extremist Organization
VTC:  Video-Teleconference